KANSAS WILDLIFE

Kansas Wildlife

TEXT BY Joseph T. Collins

WITH PHOTOGRAPHS BY Bob Gress, Gerald J. Wiens,
Suzanne L. Collins, & Joseph T.
Collins

FOREWORD BY John E. Hayes, Jr.

PUBLICATION made possible in part by
KPL Gas Service

UNIVERSITY PRESS OF KANSAS

I DEDICATE THIS BOOK to my wife, Mary Butel, who shares my interest and respect for the natural world. Her support, encouragement, and advice are reflected in the photographs on these pages. *Bob Gress*

I DEDICATE THIS BOOK to my loving and patient wife, Jan, who has assisted me in taking many of my photographs and who has encouraged me with her support of my interest in nature and wildlife photography. *Gerald J. Wiens*

I DEDICATE THIS BOOK to Chelsea, Nancy, & Randy Weaver, who care for the land so other generations may share their love of nature. *Suzanne L. Collins*

I DEDICATE THIS BOOK to my brothers, Jerry & Jeffrey, both of whom liked to catch amphibians and reptiles, and to my aunt, Norma McAllister, whose home in the country was the Garden of Eden where I saw my first snake. *Joseph T. Collins*

Published by the University Press of Kansas (Lawrence, Kansas 66049), which was organized by the Kansas Board of Regents and is operated and funded by Emporia State University, Fort Hays State University, Kansas State University, Pittsburg State University, the University of Kansas, and Wichita State University

IN A 1969 REPORT to the Secretary of the Interior, A. Starker Leopold wrote: "In a frontier community, animal life is cheap and held in low esteem. Thus it was that a frontiersman would shoot a Bison for its tongue or an Eagle for amusement . . . . But times and social values change. As our culture becomes more sophisticated and urbanized, wild animals begin to assume recreational significance at which the pioneer would have scoffed. Americans by the millions swarm out of the cities on vacation seeking a refreshing taste of wilderness, of which animal life is the living manifestation. Some come to hunt; others to look, or to photograph." It is the purpose of this book to suggest what Kansas has to offer those seeking "a refreshing taste of wilderness." As Joseph Collins makes clear, our state boasts a surprising diversity of wildlife. That diversity, that richness of animal life, depends upon our environmental vigilance.

Every Kansan has a role to play in protecting the environment. At KPL Gas Service we like to think we're setting an example for others to follow. Our company has been an industry leader in innovative methods to reduce pollution associated with electric generation. Utilizing low-sulfur coal, limestone scrubber systems, and equipment to reduce nitrogen oxide emissions, we are committed to preventing acid rain and protecting our nation's air and water. We also supply many of the homes and businesses in Kansas, Missouri, and Oklahoma with clean-burning, environmentally safe natural gas.

KPL Gas Service seeks to protect the environment in other ways. To minimize the need to build additional generating capacity, we have developed energy load management programs for our customers. We have a vegetation management program that severely limits herbicide and pesticide use. Our company developed a prototype environmental classroom at an urban Kansas City grade school. We provided land for the Dillon Natural Area in Hutchinson, and—in cooperation with the Kansas Department of Wildlife and Parks—we are developing the Jeffrey Energy Center property for multiple recreation and wildlife habitat preservation uses.

Obviously, KPL Gas Service is committed to improving life for all living things. We think that the 121 photographs that follow provide eloquent testimony why you should be too. Without that "taste of the wilderness," we would all lead duller lives. In an effort to encourage its preservation, we are pleased and proud to support the publication of *Kansas Wildlife*.

John E. Hayes, Jr.
Chairman of the Board, President and
Chief Executive Officer, KPL Gas Service

PREFACE T HIS IS A BOOK of wildlife photographs taken over the last three decades. All of the portraits in this book were made from color slides. Some, such as of birds and mammals, were taken after weeks of planning and often required elaborate set-ups involving scaffolding towers, photographic blinds, remote camera triggers, and long telephoto lens. Other animals, such as amphibians and reptiles, were captured, gently restrained, and photographed with macro lens and strobes after coaxing them to pose on carefully selected natural backgrounds.

All of us have spent enormous amounts of time traveling throughout Kansas to locate the creatures shown on the following pages. We visited every county and experienced every type of environment, endured the bitter cold of Kansas winters and the searing heat of its summer sun, rolled heavy rocks over our feet and fell into streams (always holding the camera high and dry, or almost always), took tens of thousands of photographs and discarded most, all in search of the "perfect" shot of that particular animal. Probably no more diverse collection of high-quality color slides of Kansas amphibians, reptiles, birds, and mammals has ever before been available for consideration in a single book.

During the years of hunting for these creatures, we have incurred many debts of gratitude. Thus, we take satisfaction and pleasure in thanking Ray E. Ashton, Jr., Jeffery T. Burkhart, Mary Butel, Martin B. Capron, David M. Dennis, David Edds, Connie Elpers, Joe Hartman, Connie Hay, Rex Herndon, Marilyn Holley, Carl Holmes, Errol D. Hooper, Jr., Kelly J. Irwin, Steve Jay, Tom R. Johnson, C. E. Judd, Eric Juterbock, Al Kamb, Steve Kamb, Tim Martz, Jim Mason, Gene McCauley, Larry Miller, John Moriarty, Eugene Neufeld, Thane Rogers, Stanley D. Roth, Eric M. Rundquist, Frank Smith, Travis W. Taggart, Jeffrey Whipple, Steve White, Dustin Wiens, Jan Wiens, Jennifer Wiens, the Kansas Herpetological Society, the Wichita Audubon Society, and the Wichita Department of Park and Recreation.

Chris Madson and John L. Zimmerman read and criticized the text for this book and made numerous helpful suggestions. We are indebted to them both.

Bob Gress, Gerald J. Wiens,
Suzanne L. Collins, Joseph T. Collins
25 February 1991

INTRODUCTION

IT SEEMS SUCH A SHORT TIME AGO, back in the spring of 1971, when I led the first of a series of "Saturday morning snake hunts" for children. These "hunts," sponsored by the Museum of Natural History at the University of Kansas, were actually excursions to familiarize children with reptiles and amphibians in a natural setting and were also a way of alleviating some of the fear that many youngsters (and their parents) have of reptiles, particularly snakes. Normally my classes consisted of 15 to 18 kids, aged 8 to 13, and a handful of watchful parents. We all piled on the Museum bus, the "blue goose," and took off to the wilds of Jefferson County, north of Lawrence. It was great fun, and most of the kids (and many a parent) lost much of their fear of snakes from finding and catching these creatures while an "expert" was handy to sort out the harmless ones from the dangerous types.

My notes from that time indicate that we found a lot of animals. In fact, during that April jaunt the kids and their parents, with a little help from me, found about 550 reptiles and amphibians in four hours of searching. Some of the smaller children were actually staggering under the load of wiggling wildlife in their collecting sacks. At the end of the search, after we all got back to the bus, we ceremoniously opened each bag, counted the catch, and then released all the animals back near the rocks under which we found them. The young participants had never before seen so much wildlife in one place at one time, and the experience was something they would probably never forget. A few of the adults, after seeing that many reptiles, may have experienced something they *wanted* to forget, but on the whole, they took it fairly well. None of the adults set a bad example by fainting when one of the kids lifted a big pile of snakes out of the sack and held it up for all to admire.

But the memory of that Saturday morning snake hunt made a different impression on me. After all, I had already experienced holding thirty or so wriggling Prairie Ringneck Snakes in my left hand while an outraged Red Milk Snake was chewing on the thumb of my right. The quantity of wildlife in Kansas wasn't something new to me. My wife, Suzanne, and I accompanied Larry Miller, a close friend and resident of Caldwell, Kansas, on an expedition to Harper County to find Strecker's Chorus Frogs. Upon reaching our destination near Anthony, we were stunned when thousands of these little amphibians greeted us with a lusty chorus on a cold, rainy spring night. My first trip to Cheyenne Bottoms during the fall bird migration was enough to convince me that there were plenty of birds in the Sunflower State. Seeing my first big Prairie Dog town in far western Kansas only confirmed my belief that our state had a bountiful supply of mammals, provided the habitat was there for them. What that first snake hunt did was pique my interest in the *variety* of Kansas wildlife. We actually found about twenty different kinds of amphibians and reptiles on that small hillside in northeastern Kansas. I became curious about the diverse biota of our state, because Kansas is often maligned as flat, dull, and unexciting by people ignorant of its environment. That curiosity evolved into an itch to try to observe as much Kansas wildlife as possible. I've been doing it ever since, traveling across the state in search of any new kind of critter to add to my ever-growing list. I haven't seen them all as yet, and I've been at it for a long time. Twenty years after that Saturday snake hunt the biodiversity of Kansas continues to fascinate me.

BIODIVERSITY MEANS VARIETY, or the number of different kinds of organisms in a given area, such as on Earth. To date, the world's biologists have discovered about 1.4 million living things inhabiting the Earth, and they conservatively estimate that the real total is at least 5 million and may be as high as 30 million. Biodiversity is important to Kansas and its people, as well as to the rest of the world. We cannot have a natural environment without a variety of native animals and plants, although most Kansans rarely make any connection between the presence of wildlife and their own existence. People and wildlife are, of course, related because all of us were melded by the same forces of nature that brought us to this place at this point in time. However, people have developed a complex culture, and because of it they dominate all wildlife and the natural environment. Because of that dominance, people should assume the responsibility of ensuring that we do not diminish, by even one species of plant or animal, the diversity of our planet. Why? There is no single, easy, clear-cut answer.

Some biologists argue that we can't risk destroying a species that might hold the genetic information needed to cure some dreaded disease, such as cancer. They may be right. Others assert that we don't want to risk eliminating a type of plant or animal that might provide a natural control against crop destruction. They may be right also. Still others don't want to risk abolishing life forms that might contain a key to restoring already damaged environments. And they have a point, too. All of these concerns are laudable, but are based on the hope of future revelation. Consequently, most people don't give biodiversity a high priority on their list of things to worry about.

On a more fundamental level, I think the world can little afford to allow a single species of plant or animal to disappear simply because of the unique information each life form contains. Further, at the most basic level, one of self-interest, I am concerned for our own survival. Is there a precise limit when too many species of plants and animals have been unnaturally driven to extinction? Are we teetering at that limit, a breaking point at which the Earth's environment begins to degrade so swiftly that a chain reaction sets in whereby the human species is powerless to save itself? We simply don't know the answers to these critical questions, and the current worldwide lack of interest in the natural environment and the creatures that inhabit it is cause for grave concern. But my strongest personal argument for maintaining biodiversity and all the creatures that define it is aesthetics. When I visit a Kansas environment damaged by human activities and missing its full and natural complement of plants and animals I am outraged. I feel a sense of despair, a loss of beauty, a loss of variety, a loss of choice, a loss of life. Wildlife is biodiversity. And biodiversity is the essence of a wilderness environment. It is what we seek for our souls after a hard week of work. It is what we must have to restore our sense of balance in a people-dominated world that seems more and more out of balance and at odds with Mother Nature. I consider it the responsibility of all people to act as stewards of our natural environment, but the rate of species extinction today is undeniable testimony to the failure of our governments—local, state, national, and international—to address the problem of diminishing biodiversity.

The biodiversity of Earth declines with each passing day, and the cause of it is no longer in doubt. It is now generally acknowledged by most biologists that human overpopulation is the primary threat to the biodiversity of Earth. Most biologists have known this for decades, but the subject is a sensitive one. Nonetheless, if people are serious about saving the world's wildlife, we are going to have to make it attractive to have fewer children. Each human being requires far too much of the Earth's resources, and Mother Earth has already called it quits in some areas with dense human populations. Even here in Kansas, with our blessedly small number of people, the effects of overpopulation are sharply visible in the overwhelming destruction of our prairies, rivers, and woodlands by agricultural activities driven by the desire to feed an overpopulated world at the expense of our own natural environments. If we encourage any more people to live here, we will destroy the remaining habitat for our wildlife, and this book might be the only opportunity for your great-grandchildren to glimpse the animals that once lived in our state.

SEVEN HUNDRED AND FORTY-SIX KINDS of animals live (or once lived) in Kansas. By animals, I mean vertebrates—mammals, birds, reptiles, amphibians, and fishes. I don't count insects, or any other invertebrates for that matter. Their numbers alone would soar to over 20,000 species for Kansas. To do them justice requires another, bigger book, and someone else will have to write it. This book concentrates on the terrestrial animals of our state, those creatures that are more easily seen by people afield. The 136 kinds of fishes found in Kansas are not easily observed and have been excluded. I urge you to get a copy of Joe Tomelleri and Mark Eberle's fine book, *Fishes of the Central United States* (Lawrence: University Press of Kansas, 1990) and to enjoy the color renderings therein. This book, *Kansas Wildlife,* is about creatures that spend their adult lives on land. The lone exception to this definition in Kansas is the Mudpuppy, a large, completely aquatic salamander found in rivers in the eastern part of the state. It comes pretty close to being a fish, with big red feathery gills sticking out on each side of the head and a finned tail, but its legs give it away—fishes have fins in place of limbs and can only dream of stepping up to the amphibian world.

Kansas, because of its central location on the continent, is a meeting ground for North American animals, a heartland of wildlife, and our fauna shows this diversity. Boreal animals drop down from the colder northern climes to live here, temporarily or permanently, or to pass on through to warmer places south. Some western creatures migrate east from the Rocky Mountains across our state, others slip into the southwestern corner and take up residence in our desert grasslands, and a few spread out on the High Plains and learn to live with little water. Southern wildlife pushes north into Kansas on its way back from winter quarters or settles permanently along our southern border in the Red Hills. And eastern species invade our deciduous forests, some stopping their westward movement only when they reach the Flint Hills and others continuing on until the arid High Plains prove too formidable a barrier.

Two major influences that determine the biodiversity of terrestrial animals in Kansas are precipitation and temperature. Both vary considerably from east to west and north to south, and this, plus the size and location of our state, ensures that those two phenomena will produce varied environments. Given the diverse weather, I was curious about comparing the 610 kinds of terrestrial animals found in Kansas with the diversity of our neighboring states of Colorado, Missouri, Nebraska, and Oklahoma. The figures I gathered are taken from a variety of sources, most of which were not compiled at the same time. Thus, the individual totals for each group within each state may vary by a few species either way. The following breakdown by vertebrate groups provides a detailed comparison among those five states.

NUMBERS OF ANIMAL SPECIES IN KANSAS AND FOUR NEIGHBORING STATES

	OKLAHOMA	COLORADO	KANSAS	MISSOURI	NEBRASKA
Mammals	106	123	87	74	87
Birds	442	444	429	393	410
Reptiles	83	46	64	66	47
Amphibians	51	18	30	42	14
TOTAL	682	631	610	575	558

Birds make up a preponderance of the faunas of all five states because their ability to fly allows them to invade new territory, whether temporarily (or accidently) or while migrating. For this reason they are not as affected by temperature and precipitation, because they have the option of shifting elsewhere on a seasonal basis. Further, different bird species use different areas at different seasons, and this tends to increase the species count in all the states. Thus, variation in the avifauna among the five states is low, no more than about 11.5 percent (444 species in Colorado versus 393 in Missouri). Variation is higher with mammals, reptiles, and amphibians. These three groups (with the exception of bats) are sedentary by comparison with birds and must either adapt to living the entire year in the climatic extremes of Kansas and its surrounding states, or perish. Mammals vary about 40 percent, from the montane rich fauna of Colorado with 123 species to Missouri with only 74 species. For mammals, the numerous habitats available in mountainous regions are a mecca for diversity, but not so in the more uniform environment of Missouri. Reptiles have no protective coat and experience difficulty keeping their body temperature up when the thermometer drops. They vary in numbers about 45 percent, from a high of 83 species in sunny Oklahoma to only 46 in cooler, more mountainous Colorado. Reptiles don't like the cold temperatures of high elevations in Colorado, nor the cold winters and lean rainfall to the north in Nebraska, but they thrive in the warmer, wetter weather of Kansas, Missouri, and Oklahoma. Amphibians have some concerns about temperature also, but precipitation is more critical to most of them. They need aquatic situations in which to lay their eggs and in which their larvae or tadpoles can hatch, grow, and metamorphose into adults. Variation in the amphibian fauna is high, about 73 percent, from a low of 14 species in arid, inhospitable Nebraska to a high of 51 in Oklahoma, with its low, rainy eastern ridges riddled with clear brooks and streams.

DESPITE THE GREAT VARIATION shown by amphibians, reptiles, and mammals, the 610 species of terrestrial animals found in Kansas is a good showing, falling near the middle when compared with surrounding states, which differ by 18 percent overall (682 species in Oklahoma to 558 in Nebraska). For this diversity much credit must be given to the birds. Their overwhelming numbers contribute the most to the variety of Kansas, while the mammals, reptiles, and amphibians play an important but nonetheless supporting role in enriching the number of creatures that inhabit our state. Above all, the entire vertebrate fauna of Kansas, from small and delicate Cricket Frogs to massive, ponderous Bison, from graceful Whooping Cranes to madcap Chickadees, from slow and cautious Ornate Box Turtles to exciting, high-strung Prairie Rattlesnakes, gives lie to the notion that Kansas is a land of monotony. For not only does it boast a full coterie of creatures, but also the starkly different environments that each requires to survive, and all this makes for a wilderness well worth watching, a Kansas that is treasured by us all.

Most of us yearn to visit wilderness areas and enjoy the presence of wildlife. Many of us are excited to glimpse wild creatures going about their routine of survival, but we are often disappointed because these animals are wary of us, and most kinds don't show themselves. For example, many mammals and nearly all amphibians are active at night, but most people don't wander around in the wild after dark. Also, most mammals will bite, and getting close to amphibians means going in the water and getting wet, sometimes when it is bitterly cold. Birds and reptiles are easier to see because both are active during the day. Also, birds are highly visible because they move about a lot and sing. Of all Kansas animals, it's perhaps easiest to get close to reptiles. Many can be discovered by lifting rocks, and most can be safely caught and held for closer inspection (be sure you wear stout gloves). Overall, however, wild animals normally will not sit still long enough to be photographed.

But the four photographers featured in this book have become familiar with Kansas critters over the years and have learned some techniques to get them to pose for a portrait. The animals whose images appear in this book were good enough to pause for a few seconds before the camera lens so that you might get to know them better. Look for them the next time you stroll the natural places of Kansas. If you have a hard time getting away from work and other responsibilities to take such a stroll, but still want to see Kansas wildlife, then settle down in an easy chair, or on the couch with the kids, and turn to the photographs in this book. I think the variety will surprise you as much as it surprised me back in 1971—and still does today.

KANSAS WILDLIFE

The Great Plains Toad (top right) is a big plump amphibian most abundant in the western three-fourths of Kansas, although it has invaded east along the Kansas River floodplain to the Missouri border. The best time to observe this creature in large numbers is at night around shallow pools during its spring and early summer breeding congregations. It can also be observed during rainstorms on hot summer and early fall nights as it hunts for insects on roads and highways, hopping along on the wet pavement, oblivious to oncoming traffic. Many are squashed by cars and trucks whose drivers don't see these toads. *Suzanne L. Collins and Joseph T. Collins*

Smallest of Kansas toads, the Western Green Toad (center right) is a resident of far western Kansas, along the Colorado border from Wallace and Logan counties in the north down to the Cimarron National Grasslands near Elkhart in the southwestern corner. An attractive little gem adapted to the arid High Plains, this amphibian makes use of whatever breeding habitat is available, singing and courting whenever there is rain. Because the occasion of rainfall can be a fairly restrictive time slot in the western portion of our state, this squat creature is not as seasonal as its eastern cousins and will get right to the matter of reproduction throughout spring, summer, fall, and maybe even winter if the temperatures remain warm enough and water is abundant. Western Green Toads have been designated a threatened species in Kansas. *Suzanne L. Collins and Joseph T. Collins*

The most widespread toad in Kansas, Woodhouse's Toad (bottom right) lives in great numbers in the lowlands of our state. Woodhouse's Toads consume enormous quantities of insects and are usually the amphibian found at night at the base of streetlights in our urban areas. Insects congregate around lights at night, often knocking themselves out against the glass, and the Woodhouse's Toad sits happily below, waiting for them to drop in for dinner. These animals are generally active from March to late September, but like all amphibians and reptiles, they seek deep burrows in which to survive the cold winter months. *Suzanne L. Collins and Joseph T. Collins*

This squat, pop-eyed amphibian, the Plains Spadefoot (far right), is most often found throughout the western two-thirds of Kansas, ranging east only along the Kansas River floodplain toward Missouri. To see one of these animals, grab a flashlight on a warm and stormy night and drive along roads bordering temporary pools of shallow rainwater. Spadefoots like to gather there to snare flying insects and to chorus and mate. The late Archie Carr, in his book *The Windward Road,* summed up how we feel about frogs and toads in general when he said: "I like the looks of frogs, and their outlook, and especially the way they get together in wet places on warm nights and sing about sex." *Suzanne L. Collins and Joseph T. Collins*

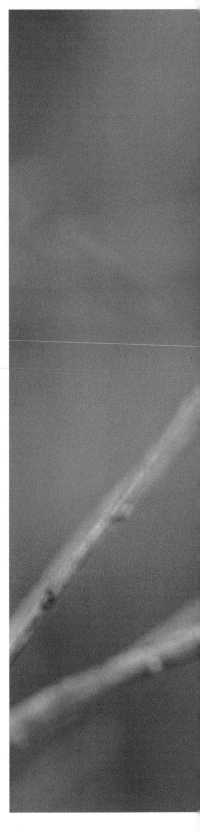

Although the Loggerhead Shrike appears at first to be an innocent gray and white songbird, this masked creature is actually a finely tuned predator with an unusual feeding technique. This bird is a carnivore, feeding on both vertebrates and invertebrates, but with a special fondness for insects. Small insects are caught and eaten on the wing, but larger prey such as grasshoppers, lizards, or mice are often impaled on a sharp object like a thorn or the barb on a barbed wire fence, sort of like a one-piece shish kebab. The Shrike, which has difficulty holding bigger prey with its small feet, then uses its strong beak to tear off pieces of its one-course dinner. This solitary species likes open country, where it is often seen perched on bushes, utility wires, or fence posts, watching for something to eat. *Bob Gress*

The Western Meadowlark is usually the bird heard calling in the background on the sound track of Hollywood western movies. Human migration and exploration across the North American continent was from east to west, so it's not surprising that this western bird wasn't discovered until 80 years after its eastern relative, the Eastern Meadowlark. Because of this delayed discovery, it was given the Latin scientific name *neglecta* in reference to the presumed "neglect" by science for overlooking the species. The Western Meadowlark is the official state bird of Kansas and is very difficult to visually distinguish from its Eastern cousin, a species also found throughout much of Kansas. The two Meadowlarks, however, have distinctive calls—the Western sings 7 to 10 variable notes in a flutelike, gurgling manner whereas the Eastern issues two clear slurred whistles which are quite musical. *Bob Gress*

Vireo is derived from the Latin word *virens,* meaning "green," and most species of vireos have some shade of green color on them. The White-eyed Vireo may not be bright green, and in fact it may require some imagination to locate this color, but just try hard and you'll see it. Vireos are very secretive and difficult to observe. The White-eyed Vireo lives in tangled thickets along the edge of forests, where its emphatically delivered song reveals its presence. It eats insects and berries during its summer visits to Kansas. In early October this bird heads south to spend the winter in an area from the Gulf of Mexico southward through Central America. *Gerald J. Wiens*

The RED FOX seems to thrive inside medium-to-large cities across Kansas. It has moved into urban areas to avoid Coyotes, which will kill and eat it. Coyotes are more afraid of people and tend to shy away from cities, and this results in Red Foxes' having a safer place to live. Red Foxes eat rabbits, squirrels, birds, fruits, and insects, and such exotic items as leftover dog food. This wild canine uses a den only when raising a litter of pups. During the rest of the year it simply roams and sleeps where it pleases. Unlike most mammal families, both the male and female Red Fox, a devoted couple thought to mate for life, rear the young together and take turns feeding and training them. *Bob Gress*

The MUSKRAT gets its name from its musky odor. This mammal likes to live in dens along riverbanks or around ponds, preferably in areas thick with cattails. It uses these plants as food and as material to build small lodges, like miniature beaver lodges. This rodent is ratlike in appearance, and its webbed hind feet make it an excellent swimmer. The fur is waterproof and insulated, allowing Muskrats to spend large amounts of time in the water. They can remain submerged for up to 17 minutes before surfacing to breathe. Muskrats are vegetarians, eating such plants as cattails, sedges, and water lilies. In some parts of our country, the meat of the Muskrat is sold as "marsh rabbit" or "marsh hare." That might be worth knowing, just in case your favorite local restaurant lists it on the menu. *Gerald J. Wiens*

The WESTERN MASSASAUGA, smallest of Kansas rattlers, also has the tiniest rattle. In fact, the rattle is so small that it is difficult to hear even at close range. So this venomous serpent has probably chalked up a fair number of unexpected people-snake encounters, particularly at Cheyenne Bottoms Wildlife Refuge, where it is very abundant. More than a few duck hunters and bird watchers who have visited the Bottoms can likely spin some Massasauga tales that are headed for legend status. But exaggeration based on an adrenalin rush aside, this reptile is actually quite beneficial, consuming many small rodents in the course of a season and keeping everyone, including the mice, alert and on their toes. *Suzanne L. Collins and Joseph T. Collins*

Big is the best way to describe the ALLIGATOR SNAPPING TURTLE. The largest freshwater turtle in North America, it reaches a maximum weight in the wild of about 315 pounds. The heaviest example from Kansas was caught back in the 1930s and weighed in at 132½ pounds, much heftier than the record for the more familiar and widespread Common Snapping Turtle, which tipped the scales at a mere 32 pounds. Alligator Snappers are confined to the streams and rivers of southeastern Kansas where they sit on the bottom and wait for prey to pass by. They probably devour anything that wanders too close, but studies elsewhere in their range show they prefer to eat other small turtles. Given the size of this creature, sitting on the bank and dangling your toes like bait in a river where Alligator Snappers live could be risky; dangling your toes from a drifting boat might best be described as trolling. *Suzanne L. Collins and Joseph T. Collins*

Although the THREE-TOED BOX TURTLE is overshadowed by the publicity received by its close cousin the Ornate Box Turtle (the official state reptile of Kansas and one of our natural symbols, thanks to the school children of Caldwell, Kansas), this creature nonetheless displays all the qualities that made its more ornate relative famous. It has a peaceful disposition, gives slow and cautious consideration to all viewpoints, and lives in a solid shell, making it a real homebody. Box turtles are the only land turtles native to Kansas, with the Three-toed restricted to the more forested areas of the eastern third of the state and the Ornate found everywhere. Both eat a wide variety of plants, berries, and fruits, as well as insects, earthworms and the like. Each season the females lay their eggs in nests beneath the ground, and their hatchlings emerge as little miniatures of mom, no bigger than a fifty-cent piece. *Suzanne L. Collins and Joseph T. Collins*

The YELLOW MUD TURTLE is a prairie animal, adapted to the open plains of western Kansas. With the exception of an isolated colony in the extreme southeastern part of the state, the Yellow Mud Turtle does not range east of the Flint Hills. It prefers quiet water, which is good because there aren't many roaring torrents in western Kansas, and it likes a habitat with a mud or sand bottom, which is also good because there's a lot of both out west. Like many western creatures, this reptile is an opportunistic predator, eating just about anything that comes along—insects, snails, crawdads, amphibians, fishes, and even aquatic vegetation. *Suzanne L. Collins and Joseph T. Collins*

Most abundant of aquatic turtles in Kansas, the WESTERN PAINTED TURTLE can be found in amazing numbers in some areas of our state. For example, in 1986 during a three-week survey at Cheyenne Bottoms Wildlife Refuge in Barton County, 703 reptiles were tallied by biologists. Of those, 288, or nearly 41 percent, were Western Painted Turtles. This colorful chelonian can be readily identified by the brilliant coral-red colors on its lower shell. No other turtle in Kansas has such a bright underside. To see underneath the critter, of course, you first have to catch it, and that is no minor feat. These animals are quite shy and plop into the water from their sunning perches at the first sign of danger. *Suzanne L. Collins and Joseph T. Collins*

In all of North America, the peak wintering area for the HARRIS' SPARROW is central and south-central Kansas. These are the largest sparrows in our nation, reaching a length of seven inches from beak to tail. They nest in Canada in a region west of Hudson Bay but head south to our area each fall, where they take up residence in hedgerows, wood margins, and brushland. With their pink bills, dark bibs, black crowns, and streaked sides, both sexes of this bird are easily recognized. Studies show that about 90 percent of their diet is seeds and fruit, with the remainder composed of insects, spiders, and snails. *Gerald J. Wiens*

The RUFOUS-SIDED TOWHEE gets its name from its call, which sounds, logically enough, like "to-whee." They also have a call that sounds like "drink-your-tea, drink-your-tea." This sparrowlike bird has a feeding behavior which involves kicking in leaves and debris to find food. Of course, it does the same thing while standing on a pile of sunflower seeds at your feeder, scattering the seeds everywhere. You can often hear this bird in the underbrush before you see it, because of the noise it makes while kicking and scratching in search of something to eat. *Gerald J. Wiens*

The GREAT HORNED OWL, often called a "hoot owl," is surrounded by mythology. It is considered by some a symbol of wisdom and knowledge because of its penetrating eyes and by others a symbol of death and disaster because it is a silent unseen predator of the night. The eyes of the Great Horned Owl are fixed forward, so in order to check out the action on the forest floor to the rear its head must swivel around for nearly 180° and look down its back. This is quite a feat and is accomplished because this owl has 14 vertebrae in its neck (people have only 7). Like all owls, it swallows small prey whole. To prevent dangerous blockage in the gut from the small bones, claws, and fur of their prey, owls regurgitate all this matter as pellets, spitting them out on the ground. The tufts of feathers on the head of the Great Horned Owl are not horns or ears, just pointy feathers that give the bird a distinctive appearance. The real ears are located below and a little behind the eyes. *Gerald J. Wiens*

The COMMON NIGHTHAWK has a catchy name—though deceptive—because this bird is not a hawk and doesn't spend much time cruising around in the dead of night. Nighthawks are common all over Kansas in summer and are active in the dim light of dusk and dawn, chasing insects on the wing. Males exhibit a fascinating habit during courtship of first hovering high in the sky, then tucking their wings to their bodies and diving earthward. Gathering speed in the dive, suddenly they open their wings in a swoop that creates a startling "roar" as the air rushes by, a delight to observe and hear. The Common Nighthawk belongs to a family of plain-colored birds called "Goatsuckers," which includes the Whip-poor-will. The plumage of the Nighthawk enables this bird to camouflage itself while incubating its eggs, which are laid on the ground. *Bob Gress*

The Broadhead Skink is designated a threatened species in Kansas because people are continually destroying its woodland habitat for yet another civic "improvement." Broadhead Skinks are restricted in our state to the forests along the eastern border and prefer to live in large dead trees that are standing or leaning near water, particularly swamps and marshes. These big lizards are notorious for using abandoned woodpecker cavities as places to hole up and nest. Males have striking orange heads, brightest during the spring breeding season. Females, such as the one in this photograph, are less colorful. *Suzanne L. Collins and Joseph T. Collins*

The Western Slender Glass Lizard isn't a snake, it just looks like one. Actually, it's a lizard without legs that gets all broken up over being grabbed or trod on—literally, sometimes into several pieces, each wiggling and squirming and writhing while you stand there and worry about what grave injury you may have done to the creature. But the joke is on you. That wiggling thing you are staring at is just a piece of tail, and the main body of the Western Slender Glass Lizard is probably long gone from the place of your encounter. These reptiles are just about all tail—at least two-thirds of their length—so they have a lot to give in order to escape with their important parts. They eventually grow a new tail, usually shorter and less colorful than the original. *Suzanne L. Collins and Joseph T. Collins*

The Northern Prairie Skink, a colorful and nervous little lizard, and its other Kansas relative, the Southern Prairie Skink, are distinctive among North American members of their group in having a range that is restricted to a narrow corridor from southern Manitoba south down through Kansas to the Gulf Coast of Texas. Most other skinks on this continent live in the east or the west, but the Prairie Skink claims the middle of the nation as its homeland. Like most lizards, it eats insects and other terrestrial invertebrates and spends much of its time beneath rocks and logs. To see one, you have to look in those places, but be careful. Other creatures live under rocks and logs also. *Suzanne L. Collins and Joseph T. Collins*

As its name implies, the Prairie Racerunner is the fastest lizard in Kansas—try to catch one in August at high noon, when the Kansas sun has warmed it up and your presence has given it an adrenalin rush. It's a good way to suffer heat exhaustion, but go ahead and try it. After you tire of the chase (you will surely lose), retire to the shade and wait for dusk. Then find the nearest rocky hillside and flip over the rocks in the cool of the evening. You'll find that the Prairie Racerunners are colder and very sluggish when the sun is down— and much easier to catch and observe. *Suzanne L. Collins and Joseph T. Collins*

The young EASTERN SCREECH-OWLS in this photograph are a typical gray color, but a reddish-colored variant may also be encountered in this species. Screech-owls don't screech but emit a soft, mournful, quavering call that is most unbirdlike. This small owl, generally only 7 to 10 inches tall, likes to live in natural tree cavities and woodpecker holes. Great Horned and other big owls catch and eat Screech-owls, so many of these little owls have overcome their fear of people and settled in or near towns to avoid their bigger predatory cousins. This urban living seems agreeable to the Screech-owl, and it is often seen making late night use of bird baths and feeding on insects attracted to street and porch lights. It also consumes mice and other creatures active at night. *Bob Gress*

The EASTERN COTTONTAIL breeds and breeds and breeds from February to September, usually producing 3 to 4 litters during that time span. Each litter is usually 4 to 5 young, with the potential to produce 9. Within hours after giving birth, the female mates again. Assuming all young survived, a single pair could generate 350,000 rabbits in 5 years. Fortunately, the death rate for this small mammal closely matches its birth rate (very few Eastern Cottontails live more than a year), or we would otherwise be overrun with the little beasts. This rabbit can leap 10 to 15 feet and often will stand on its hind feet to observe the surroundings. When pursued by an enemy, Cottontails circle their territories and hop or jump sideways to break their scent trail. They dislike getting wet, but will swim if necessary to escape predators. *Gerald J. Wiens*

BLACK-TAILED JACKRABBITS and jackrabbits in general were once exceedingly abundant in western Kansas, but their numbers appear to be dwindling, although population fluctuations in this animal are known to be extreme. Jackrabbits are well built for survival on the Great Plains, with large ears and eyes for detecting approaching predators. Unlike the Eastern Cottontail, which most often runs a short distance to a hiding place such as a brush pile, Jackrabbits are not inclined to hide, but escape by running and continuing to run — they just keep on going! And with good reason, for the enemies of this hare are legion and include Coyotes, Foxes, Bobcats, and Golden Eagles. These big hares have a home range of 1 or 2 square miles and can reach speeds of 30 to 35 mph. Given a good scare, they can jump 15 to 20 feet. *Bob Gress*

Also known as "groundhog" or "marmot," the stubby Wood-chuck, according to biologists, is a squirrel, albeit one with a short bushy tail and a dislike of high places. Actually, it sometimes climbs small trees in search of tasty greens to eat or to stretch out and take the sun, but when danger is near it heads for the opening to its underground burrow. Lots of bigger animals like to eat this chubby mammal, but once it dives into its underground chambers there is little hope of catching it. Woodchucks are residents of eastern Kansas where they inhabit meadows and woodlands, occasionally making a loud shrill whistle if a predator is near or something is not to their liking. This is one of the few mammals in Kansas that is a true hibernator. During hibernation its body temperature falls from about 97°F to less than 40°F, its breathing slows to once every six minutes, and its heartbeat drops from over 100 to only 4 per minute. *Gerald J. Wiens*

A member of the squirrel family, the Thirteen-lined Ground Squirrel is a close relative of Prairie Dogs and Woodchucks. Its striped and spotted coat is not merely decoration but provides a pattern that blends with sunny and shady spots in grassy areas. Originally found only in short-grass prairies, this small mammal can now be seen along roadsides and in yards, cemeteries, golf courses, city parks, and anywhere else that grass is kept short. This is also one of the few Kansas mammals that undergoes true hibernation, lowering its body temperature to around 37°F in its winter den. After hibernating for up to 5 months, this little creature emerges from its burrow in spring to hunt for seeds, berries, nuts, insects, and sometimes even mice or shrews. Like the Black-tailed Prairie Dog, these squirrels will stand on their hind feet to better observe their surroundings. *Bob Gress*

The Wild Turkey was once abundant in Kansas but was extirpated in the early 1900s. They were re-established in the 1950s and are now common in the riparian woodlands of our state. The bird in the photograph is an adult male strutting his stuff and displaying the red caruncles extending down from the area of the beak. Below that is the beard, which is nothing more than a collection of feathers, the function of which is unclear. Wild Turkeys are very alert and intelligent birds when compared to their domesticated cousins. Given the ultimate demise of the barnyard version, one would think it would smarten up. With the oven as an alternative, it seems a tame turkey headed for the Thanksgiving feast would have enough incentive to match up intellectually with its brethren in the wilderness, and figure out how to avoid being the table centerpiece. *Bob Gress*

The Ring-necked Pheasant is not native to the Western Hemisphere but was introduced to Kansas from Asia in the late 1880s to provide hunters with a recreational species that could live in cropland. This bird is now well established over most of North America where suitable climatic conditions and agricultural fields exist. Pheasants are hunted extensively and provide many Kansans with hours of exercise and sport. Although the bright, gaudy colors don't provide this predator-wary bird with good camouflage, the males survive due to an alert nature and the habit of flushing wildly at the first sign of danger. *Bob Gress*

An encounter with a WESTERN HOGNOSE SNAKE (top right) is an experience. When frightened by something as big as a human being, it hisses, spreads a hood, and lunges at the intruder. If such aggressive behavior fails to drive you away, this snake curls up on its back and opens its mouth in agony, writhing as if wounded, like the snake in this photograph. Finally, it relaxes and "plays dead." However, careful and quiet observation from a distance will reveal that the snake was waiting for you to lose interest and leave, at which time it will right itself and quickly wiggle off to safety. Unfortunately, the initial aggressive displays by this harmless little reptile often result in its winding up truly dead, the victim of ignorance and an irrational fear of all snakes. Western Hognose Snakes are gentle bluffers and deserve better than they often get from people. *Suzanne L. Collins and Joseph T. Collins*

The PRAIRIE RINGNECK SNAKE (center right) is the most abundant snake in eastern Kansas, with estimates of up to 750 per acre in optimal habitat. That's probably too many snakes for most folks, but few Kansans will see any of these small harmless snakes. They are very secretive and spend most of their life under rocks or beneath the ground looking for earthworms, their favorite meal. The Prairie Ringneck Snake has an interesting defense posture when bothered by an enemy—if poked, bitten, or pinched, it tucks its head beneath its body coils and curls its tail tightly, lifting it high in the air to expose the bright red underside. This is called flashing, an activity apparently designed to startle or distract an intruder away from the head and give the snake more time to escape to safety. *Suzanne L. Collins and Joseph T. Collins*

The EASTERN YELLOWBELLY RACER (bottom right) is a fast and aggressive snake, so aggressive that during the spring courting season it has been known to approach or follow people who wander into its territory. But the Eastern Yellowbelly Racer is harmless and usually will retreat if approached too closely. These reptiles reach a maximum length of nearly six feet and prefer to live in open areas, particularly around wetlands, where they hunt for mice, frogs, and other snakes. *Suzanne L. Collins and Joseph T. Collins*

Most delicate and graceful of Kansas snakes, the ROUGH GREEN SNAKE (far right) gets the name "rough" from the longitudinal ridge or keel along the center of each of the upper body scales. An inhabitant of the eastern part of our state, this slender reptile spends its time in bushes and trees near streams and lakes, waiting motionless for unwary insects to land close enough to provide lunch. Rough Greens are completely harmless and inoffensive. They never attempt to bite, relying totally on their cryptic coloration to avoid trouble. *Suzanne L. Collins and Joseph T. Collins*

RED-HEADED WOODPECKERS live mostly in open deciduous woodlands but are often seen far from trees out in the open prairie, flitting from fence post to utility pole to windmill. The male and female in this photograph look very much alike and are difficult to distinguish. As fall approaches, these flashy-looking birds spend part of their time storing acorns and pecans for food in winter. This bird likes to live in tree cavities but will also drill holes in telephone poles. It can often be seen darting from its perch in chase of insects. *Gerald J. Wiens*

The RED-BELLIED WOODPECKER
is so named because of the very
pale reddish or pink belly. The
color, however, is rarely seen because the woodpecker is usually
belly-up against a tree or branch.
It is sometimes called the "zebra"
or "ladderback" woodpecker because of the transverse black and
white bars that cross its back and
wings. The staccato pounding of
its bill is muted when feeding but
loud and ringing when this bird is
signaling. The seemingly unneces-
sary drumming and hammering
on stovepipes and metal gutters is
this woodpecker's method of ad-
vertising its claim to a territory.
Gerald J. Wiens

Standing motionless in the shallow waters of Kansas lakes and ponds, the stately GREAT BLUE HERON patiently waits for unsuspecting prey to come near its sharp, pointed bill. Great Blue Herons are the most widespread and best known of North American herons and are found throughout Kansas. They feed primarily on fishes but will also consume insects, frogs, snakes, small birds, and small mammals. In Kansas, they seem to show a preference for building their nests in colonies in tall sycamore or cottonwood trees along rivers and streams. *Gerald J. Wiens*

The SNOWY EGRET often builds its nest in trees and, despite its awkward appearance, is quite graceful as it moves about the treetops. It is sometimes called "yellow slippers" because its yellow feet are in striking contrast to its lanky black legs. The Snowy Egret was driven to the edge of extinction in the late 1800s by hunters in pursuit of its beautiful breeding plumes, or aigrettes, which were fashionable in the millinery trade for women's hats and from which this bird gets its name, egret. Through protection provided by the federal government, this showy bird has made a comeback and has now expanded its range into Kansas. *Bob Gress*

The PRAIRIE VOLE is one of the most prolific mammals known. A family of these little rodents was observed to have 17 litters in a single year, each litter consisting of 4 or 5 young. Females can bear young before 7 weeks of age. The populations of this rodent vary—in bad years it may be as low as 15 to 40 per acre; in good years 60 to 250 per acre. The Prairie Vole is an important food source for many other creatures such as Bullsnakes, Red-tailed Hawks, Bobcats, Coyotes, and other predators. Prairie Voles build many runways beneath the grass in order to move about quickly without being seen by predators. Most of the time this is a very successful mode of travel, but every now and then a Coyote gets lucky. *Bob Gress*

In a survey done a few years ago, the COYOTE emerged as the only Kansas animal that ranked both in the "most favorite" top 10 and the "least favorite" top 10. People have poisoned it, trapped it, gassed it, and hunted it on land and from airplanes. The Coyote has not only survived this onslaught, but it has greatly increased its range in the face of well-financed persecution. Coyotes were probably once limited to the western mountains and the Great Plains, but because people destroyed their major competitor, the Gray Wolf, they have now expanded into eastern North America. Cunning, tenacious, and resourceful, the Coyote is the symbol of a survivor. Its howl is one of the stirring sounds of the prairies, and we would be less without it. *Gerald J. Wiens*

If alarmed, a WHITE-TAILED DEER will raise or "flag" its tail, exhibiting a large bright flash of white. This "high-tailing" communicates danger to other deer and helps a fawn follow its mother in flight from enemies. The spots on the fawn in this photograph help to camouflage it when resting or bedding down in wooded areas. Deer are good swimmers and have a top running speed of 35 mph. When nervous, a White-tailed Deer snorts through its nose and stomps its feet, warning other nearby deer of danger. Contrary to popular belief, the age of a White-tailed Deer cannot be told by the tines on its antlers. The age of younger White-tails can be told by the wear on their teeth; older adults can be aged by examination of tooth roots. Neither technique works well on a living deer. *Gerald J. Wiens*

Even though the U.S. Congress outlawed the shooting of the BALD EAGLE in 1940, our national symbol still declined in numbers through the 1950s and 1960s. Once it was discovered that a pesticide was contaminating the fishes that these eagles ate and was interfering with the proper development of eagle eggs, our government banned the offending chemical. Eagles are now making a comeback and are most easy to observe in Kansas during winter, roosting along rivers and around large reservoirs where they soar and glide, ever watchful for an unwary fish or duck to eat. The Bald Eagle is still designated an endangered species nationwide. For the first time in the recent history of Kansas, they were observed in 1989 nesting and raising their young at Clinton Reservoir near Lawrence. *Gerald J. Wiens*

The COOPER'S HAWK is named for William Cooper of New York, who collected the first examples of the species in the early 1800s. This raptor has short stocky wings and a relatively long tail when compared to most other hawks. The Cooper's Hawk has a habit of alternately flapping and gliding while flying and tends to live in woodlands. It is about the size of a crow, with a wingspan of 27 to 36 inches. When hunting, it glides and maneuvers through woodlands in low swift flight. This species sometimes attacks poultry and may deserve the name "chicken hawk" more than most other hawks. Cooper's Hawks primarily eat small mammals and birds and are present in Kansas mostly during the winter months.

Gerald J. Wiens

The WESTERN RIBBON SNAKE (top right) is a slim and graceful snake that lives around marshes, swamps, ponds, roadside ditches, and small lakes. Poised motionless near the water like a reed or stalk, it alertly watches for any movement by small frogs and fishes, then glides swiftly to capture such tasty morsels. Ribbon Snakes are generally found throughout Kansas but are scarce on the High Plains, where there is limited permanent water. Each year female Western Ribbon Snakes give birth to litters of up to 27 young, each of which are pencil lead–thin versions of their parents. *Suzanne L. Collins and Joseph T. Collins*

Most Kansans don't think of garter snakes as big, but the RED-SIDED GARTER SNAKE (bottom right) reaches a maximum length of about 4 feet. When angered, it's a fairly impressive creature, curling in a defensive posture and flattening its coils to make its body mass look larger. This is the only reddish garter snake in our state. Like most other garter snakes, it prefers to live near marshes and wetlands, where there is an abundance of good things to eat—earthworms, snails, insects, minnows, and frogs. Females of this species produce enormous litters, with one female known to have given birth to 80 young. Few baby Red-sided Garter Snakes survive because they provide so much food for many other creatures that live in our state. *Suzanne L. Collins and Joseph T. Collins*

Grassland deserts are home to the CHECKERED GARTER SNAKE. It is found in Kansas only along the southern border of our state, from Sumner County in the east to the Cimarron National Grasslands in the extreme southwestern corner. Unlike most other garter snakes, this reptile strays good distances from water, apparently being much less dependent on aquatic habitats as sources of food. The Checkered Garter Snake is not a picky eater and consumes whatever it finds, from lizards and mice to toads and insects. It reaches a maximum length of about 42 inches and has been designated a threatened species in Kansas. *Suzanne L. Collins and Joseph T. Collins*

The Beaver (left) is the largest North American rodent and is characterized by its large flat tail, webbed hind feet, and well-developed front teeth. This mammal combs its fur with small grooming claws on its hind feet which are also used to apply castoreum, an oily secretion from a scent gland near the base of the tail. The engineering skills of this aquatic mammal are legendary. Beaver dams give diversity to streams and rivers and act as natural checks to slow down the spring runoff from rain. The pools behind these dams provide resident Beavers with quick escape routes as well as an easy means of floating logs and branches to the dinner table. Kansas Beavers generally dig their dens in stream banks rather than building the classic freestanding lodges common in the western mountains and Canada. *Gerald J. Wiens*

The tiny Western Harvest Mouse (above left) is a very good climber, ascending grass stems and shrubs to harvest seeds, its preferred food. Fortunately, these rodents do this only at night, a good strategy when you consider how exposed a mouse might be on the end of an upright grass stem in broad daylight, with big predatory hawks cruising the skies also looking for lunch. These mice usually nest on the ground, but sometimes build a little round nest located 3 to 4 feet above the ground. The Western Harvest Mouse is a fastidious creature, spending much time cleaning itself. *Bob Gress*

A small rodent found in the western half of Kansas, the Northern Grasshopper Mouse (above right) is a predator, hunting and catching insects such as grasshoppers, which it eats with relish. It will also kill and consume other mice, tracking them down with its excellent sense of smell. After dispatching another rodent, these tiny mice will sometimes stand on their hind legs and emit a high-pitched squeaking howl, sort of like a miniature wolf. They also use their vocal signals to establish territories. Unlike other mice, they have a strong personal odor, possibly from their habit of eating the flesh of other mice. *Bob Gress*

Until very recently, the elusive COMMON MAP TURTLE was thought by many to be extirpated in Kansas, a victim of pollution because of poor agricultural practices. Except for one example found in 1952, it had not been seen again until the summer of 1990 when a group of biologists from Emporia State University made a search and found it—in fact, they found quite a few—much to the delight of environmentalists. Apparently this turtle was adept at staying out of sight and did so for decades, floating peacefully on isolated stretches of small rivers and streams in eastern Kansas and lazily munching its favorite foods—freshwater mussels and other crustaceans. *Suzanne L. Collins and Joseph T. Collins*

At one time it was thought that every hawk was a "chicken hawk," and for many farmers the only good hawk was a dead one. Because the RED-TAILED HAWK is the most common hawk in Kansas during any season of the year, many of these birds were killed to protect poultry flocks and songbirds. Despite its name, the tail feathers of this big bird are never a true red, but rather a reddish orange or rufous. During the first year of life its tail is brownish with faint narrow bars. One of the best identification characters for this bird is the presence of a dark band across the belly. The band may be solid or consist of a few dark spots, as in this photograph. This raptor is most commonly seen perched on top of isolated trees, utility poles, or fence posts along highways. It has acute vision, 9 times more powerful than ours, and its eyes are capable of rapidly adjusting from wide view to an extreme closeup, much the same as zoom binoculars. Red-tailed Hawks eat a wide variety of food, from small insects to fair-sized mammals, but a study in northeastern Kansas has shown that snakes (not chickens!) make up nearly 50 percent of the diet of these big birds. *Gerald J. Wiens*

The AMERICAN KESTREL is often called a "sparrow hawk," but this name is inappropriate because its diet is far more varied than just small birds. Insects are the dietary preference in summer; mice and small birds are consumed throughout the year. This graceful raptor, a member of the falcon family, is the smallest of all North American hawks and can be commonly observed perched on fence posts and telephone poles along Kansas roads and highways, alert for any sign of prey below. Its eyes contain minute yellow oil droplets that act as a filter to shut out haze and glare. Combined with the fact that its eyes are as powerful as a pair of eight-power binoculars, it is no wonder that the American Kestrel has a good command of all it surveys and plays havoc with the mice and other small creatures that cross its vision. *Gerald J. Wiens*

Historically one of the most widespread species of birds, having lived on all continents except Antarctica, the PEREGRINE FALCON has been dramatically reduced in numbers worldwide due to pesticide use. However, it has made a significant comeback in the United States due to captive breeding programs and a decrease in the use of certain pesticides. Captive-raised birds have been released in urban areas where they nest in tall buildings and hunt pigeons. Our federal government has designated it an endangered species. In its hunting dives for prey—usually other birds—the Peregrine Falcon can reach speeds over 100 mph. *Gerald J. Wiens*

The habits of the LESSER PRAIRIE-CHICKEN are similar to those of the Greater Prairie-chicken. In Kansas, these birds occupy entirely different portions of our state. Whereas the Greater prefers the tallgrass prairie, the Lesser is found a little farther to the south and west in the sand-sage prairie. Telling these two apart can be vexing, but basically the Lesser is a paler version of its cousin and has reddish, almost plum-colored, neck sacs. The Greater and Lesser designations aren't much help as characteristics either. The Greater grows to a total length of 18 inches, whereas the Lesser grows to only a puny 16 inches—a difference which loses much of its significance when a short Greater sidles up to a tall Lesser. Trying to get close enough to measure either species will occupy a lot of time, not to mention the excitement it will give the birds. The Lesser Prairie-chicken is a prairie treasure, and Kansas boasts a larger number of these birds than any other state. *Bob Gress*

The GREATER PRAIRIE-CHICKEN congregates each spring at "leks" or "booming" grounds where the male puts on a spectacular display. The courtship dance begins with each male rapidly stomping the ground with his feet while erecting the feathers on each side of his neck and pointing his tailfeathers skyward. The neck begins to distend as the male inflates orange-colored sacs on each side of his throat, using them to resonate a booming sound. All this is done to impress the females, who spend most of their time wandering around ignoring the males. Despite this lack of interest by the females, Kansas has the largest known populations of Greater Prairie-chickens in the world, with a current estimate of about 750,000 birds. This is due primarily to the natural grassland habitat of the Flint Hills, which provides optimal living areas for these birds, but is also because the females eventually do pay attention to the males. *Gerald J. Wiens*

Given its name because of its mournful song, the MOURNING DOVE has a large crop, which allows it to descend to the ground, eat and store a large amount of food in a short period of time, and quickly beat a retreat to safer (higher) environs. This bird has rather loose plumage, a characteristic that gives it an edge in escaping predators, who may lunge for a dove only to come away with a mouthful of feathers. Unlike most birds, Mourning Doves feed their young "pigeon milk," a high-fat, high-protein cheesy substance secreted from the lining of their crop. This is one of the few birds that can "suck up" drinking water, rather than "bill it up" like other birds. *Bob Gress*

The YELLOW-BILLED CUCKOO is sometimes dubbed the "rain crow" because it is thought to call more frequently just prior to a thunderstorm. This bird is a summer resident of Kansas, spending its winters in the warmer parts of South America. It is a shy, quiet bird that tends to remain concealed in the foliage of trees and shrubs, where it devotes its efforts to catching and eating great numbers of hairy caterpillars, as well as other insects, fruits, berries, treefrogs, and small lizards. This species constructs a somewhat flimsy nest when it comes time to lay its eggs. The nest often resembles that of the Mourning Dove; but the eggs, usually two or three, are a pale blue-green instead of the Dove's glossy white. Every now and then the female Cuckoo lays her own eggs in the nests of other kinds of birds. This deceptive act relieves the Cuckoo parents of the task of raising their young from hatchlings through the ever-difficult time of puberty. Meanwhile, the unsuspecting owners of the nest try to figure out where all the eggs came from and how they will make ends meet. *Gerald J. Wiens*

A creature of legend around the scout campfire, the COMMON SNIPE has probably been actively, if not misguidedly, sought by more Girl Scouts and Boy Scouts than any other creature in the world. Many scouts end up believing the Snipe doesn't exist. But yes, scouts, there is a Snipe. Furthermore, it is one of only two shorebirds in Kansas which can be legally hunted. The Common Snipe is abundant in wet boggy places where it probes the mud with its long bill in search of invertebrates. Well camouflaged, the secretive Snipe is seldom seen before startling an intruder by bursting into a zig-zag flight and zooming off to safety. *Bob Gress*

MINK are found throughout Kansas, but are least abundant in the western part of the state. They prefer to live around water, particularly in brushy areas. A relative of skunks and badgers, this small mammal is active mainly at dusk, during the night, and at dawn, spending that time in search of food. It is a predator, and although its prey depends on the season, it is known to consume small mammals, crayfishes, insects, birds, amphibians, reptiles, and fishes. The Mink usually dens under tree roots along a riverbank, or in tree holes or hollow logs. Sometimes it will live in abandoned beaver or muskrat dens. Males and females maintain separate dens and are essentially solitary. The fur of this little animal is prized for making coats, but the Mink probably feels there are other materials better suited for this purpose. *Bob Gress*

The EASTERN SPOTTED SKUNK is much smaller, faster, and more agile than its close cousin, the larger and chubbier Striped Skunk, and a good tree climber to boot. Its spraying behavior is fascinating. When confronted by an enemy, this smelly little animal raises its tail in a threat posture. If the intruder refuses to retreat, the skunk stands on its forefeet, raises its tail again, and sprays its foul musk with amazing accuracy for a distance of up to 12 feet. This is extremely effective against ground-level foes, but its chief predator, the Great Horned Owl, can strike unseen from above and take a baby skunk before its mother can spray in defense of the youngster. Because of environmental destruction, the Eastern Spotted Skunk has been designated a threatened species in Kansas. *Gerald J. Wiens*

The AMERICAN WHITE PELICAN does not breed in Kansas but migrates through during spring and fall and sometimes stays at Cheyenne Bottoms Wildlife Refuge during the summer. A flock of White Pelicans in flight is a joy to watch, spiraling gently upward on thermals with a minimum of effort and then gliding off to fishing areas. While flying low over water they often cruise in single file, undulating in unison as if riding some unseen wave. These birds use their large pouches for fishing, shoveling it beneath the water like a bucket to scoop up their favorite food. Pelicans are very large birds, with a maximum wingspan of 9 feet and a weight of up to 17 pounds. In comparison, eagles generally do not exceed 12 pounds. *Bob Gress*

Its brilliant red eyes give the EARED GREBE a distinctive appearance and make it easy to identify with good binoculars. This bird can be seen regularly in Kansas during its spring and fall migrations and has been known to nest at both Quivira and Cheyenne Bottoms Wildlife refuges. Grebe nests consist of floating platforms constructed of aquatic vegetation. Eared Grebes have the ability to float halfway beneath the water, with only their heads and necks showing, and are excellent underwater swimmers. This bird feeds primarily on aquatic insects, tadpoles, frogs, small fishes, and some water plants. *Bob Gress*

71

The EASTERN CHIPMUNK is an animal of eastern Kansas. Although mostly a ground-dweller, it will often climb trees to escape enemies or search for foods such as seeds, nuts, berries, and fruits. Chipmunks have cheek pouches which can be greatly expanded to hold good things to eat. When communicating, they make a high pitched "chipping" noise which sounds like a bird. Unlike cute cartoon chipmunks, the adult Eastern Chipmunk with its large front teeth can give a solid nasty munch. So don't pick one up—just feed it some nuts and leave it alone. *Bob Gress*

The CAROLINA WREN is the largest of the wrens found in the eastern United States. The resourceful bird in the photograph built a nest in a hanging flower pot on a back porch. Like other wrens, the Carolina Wren does not undergo a seasonal change in its plumage, nor do its young differ much from their parents. This petite bird has a beautiful song, often described as "tea kettle, tea kettle, tea kettle," and sung in the same tone as the call of a Cardinal, a species with whose call it is sometimes confused. During a severe winter in the late 1970s, the population of these birds in our country plummeted. In recent years they have made quite a comeback, both in Kansas and elsewhere. *Gerald J. Wiens*

Early American colonists gave the orange-breasted AMERICAN ROBIN its name because it reminded them of the Robin Redbreast of their European homeland. Both American and European robins are members of the thrush family, and the young of both exhibit a speckled breast. The northward spring migration of the Robin follows very closely the average daily temperature of 37°F. Inexplicably, certain Robins consistently migrate each year, while others always choose to hang out in Kansas all winter long. Recently, a series of experiments has shown, contrary to popular belief, that these birds do not find earthworms by sound, but instead rely on sight to locate their lunch. So the next time you see one of these birds cock its head to the ground, realize it is using its eyes, not its ability to hear. *Gerald J. Wiens*

The Virginia Opossum is the only Kansas mammal with opposable thumbs (on its hind feet) and fifty teeth (more than any other Kansas mammal). It also has the dubious distinction of possessing the smallest brain in proportion to body size of any mammal in our state. Despite this, it survives very well because it can raise up to 13 young per litter. It is our only mammal with a prehensile tail, which it uses for balance and security when climbing around in brushpiles and trees. *Gerald J. Wiens*

The eyes of Eastern Moles are nonfunctional, able to distinguish only between light and dark. But they don't need them. Moles have a sensitive nose and excellent hearing, and these, plus their ability to pick up ground vibrations, make it difficult to surprise one of these creatures. They emit a musky odor from a scent gland on the belly that is distasteful to predators. Mole tunnels are primarily feeding tunnels where these mammals prey mainly on earthworms by tracking them with a very sensitive nose. *Gerald J. Wiens*

Most people think spring starts on the twentieth of March. Not so. The WESTERN CHORUS FROG (far left), most widespread and abundant of chorus frogs in Kansas, is one of the earliest creatures in our state to announce the onset of the new season. With the first rains of late January or February, it emerges from its snug quarters beneath the ground. Excited by the prospect of courtship and breeding, it hops to the nearest pool of water and begins to chorus, announcing to all who listen that spring has arrived. *Suzanne L. Collins and Joseph T. Collins*

The tiny NORTHERN SPRING PEEPER (top left) was first discovered on the North American continent at Fort Leavenworth, Kansas, and, on the basis of that find, described as new to science in 1838. But the Peeper has never again been observed at the Leavenworth military reservation, posing a real mystery for biologists. Maybe it was a mistake—maybe some frontier soldier found or observed that first Peeper in eastern Missouri and brought his catch to the Kansas fort, not bothering to mention its eastern origin. The frog was eventually located at two other areas in our state, both to the south along the Missouri border, but today this small amphibian is restricted to a small series of marshy pools in Kansas, in the Ozarkian area of Cherokee County. Last we heard, it was barely holding its own. It needs some protection—and soon—or it may disappear from our environment. Spring Peepers have been designated a threatened species in Kansas. *Suzanne L. Collins and Joseph T. Collins*

Abundant throughout the prairies of south-central Kansas, the small attractive SPOTTED CHORUS FROG (center left) emerges from its winter haunts each spring when seasonal thunderstorms fill low areas with water and form the life-giving pools and marshes so necessary for these creatures to reproduce. Unlike reptiles, but much like their closer cousins, the fishes, all Kansas frogs, toads, and salamanders lay their eggs in or very near water so the eggs won't dry up. After hatching, the gilled larvae stay moist and healthy in their watery cradles until time for them to metamorphose into miniature adults and face the rigors of life on land. *Suzanne L. Collins and Joseph T. Collins*

Tiniest of Kansas frogs, BLANCHARD'S CRICKET FROG (bottom left) grows to a maximum length of only 1½ inches. As a result, this amphibian consumes only the smallest of insects. The call of this creature is very distinct, sounding like a bag of marbles being vigorously shaken. Blanchard's Cricket Frog is found throughout our state but is least common on the arid western High Plains. Its abundance in some areas makes it an important food item for other animals, such as Ribbon Snakes, shorebirds, and small mammals, as well as bigger kinds of frogs. *Suzanne L. Collins and Joseph T. Collins*

Like its cousin, the Yellowhead, the RED-WINGED BLACKBIRD thrives around marshes and wetlands, particularly those with good stands of cattails. Anyone who has ever gone fishing is familiar with the song of this bird. Red-winged Blackbirds are found throughout Kansas and get their name from the bright red patch or epaulet on each shoulder of the male. These patches are most distinctive during the spring courtship and are less conspicuous at other times of the year. This bird is quite gregarious in winter, traveling and roosting in large flocks. It dines on insects, grains, and seeds. *Bob Gress*

The YELLOW-HEADED BLACKBIRD is a warm-weather visitor to Kansas, often seen in large numbers at Cheyenne Bottoms Wildlife Refuge during the summer months. Its rasping song resembles a rusty-hinged barn door being opened. Courtship by the male involves contorted displays as he pursues the female. The Yellow-headed Blackbird breeds in colonies sometimes numbering thousands of individuals and can be approached closely. Large flocks forage in fields and open uplands for seeds, grain, fruits, and insects. This bird winters in the southwestern United States and Mexico. *Bob Gress*

Sometimes given the misnomer "mountain boomer" by folks in Oklahoma, the EASTERN COLLARED LIZARD makes no sound and doesn't much care for high mountains, which are usually too cold for most reptiles. The "boomer" moniker probably came about when an Okie saw one of these lizards perched atop a big boulder while at the same time some other creature hidden nearby made a "booming" noise. Thus are legends (and fallacies) born. The Eastern Collared Lizard may be silent, but it has an impressive outsized head and the startling habit of running only on its hind feet when threatened by an enemy. It picks up its front feet, tucks them against its chest, and takes off. Because these lizards like rocky habitats, they are very difficult to catch. On the other hand, their speed and agility stand them in good stead when they search for their favorite food, other small lizards, which they swallow with a couple of gulps. *Suzanne L. Collins and Joseph T. Collins*

The TEXAS HORNED LIZARD is a prickly looking reptile with long, sharp spines protruding from the back of its head. The color pattern of its rough skin enables it to blend well in sandy or gravelly areas, making it difficult to see while motionless. When active it likes to run around in the noonday sun in August, out in the sandy, open areas of Kansas. With that lifestyle, plus its formidable spines and cryptic skin, the Texas Horned Lizard has few enemies. It does have a peculiar habit. Sometimes, when picked up and held, it will squirt small jets of blood from the corners of its eyes. Biologists used to think this was a defense measure and that the blood was distasteful to predators. We now know this squirting is accidental, apparently the result of either stress or slight pressure while handling the lizard. The lizard doesn't appear to suffer any ill effects from the loss of blood. But if you catch one, try not to squeeze it too hard. *Suzanne L. Collins and Joseph T. Collins*

The male PURPLE FINCH isn't really purple, but more of a raspberry red. Females exhibit none of the male's reddish coloration, but are instead streaked with brown and white, resembling many of the sparrows. Both sexes exhibit short thick bills, an essential tool for cracking open the tough husks of seeds. This bird lives in groves and orchards, dining on such seeds as well as fruit and insects. Purple Finches are winter visitors to eastern Kansas, and with the arrival of spring, they return north to their Canadian nesting areas. *Bob Gress*

Like most members of the finch family, the AMERICAN GOLDFINCH has a cone-shaped bill adapted for eating seeds, and the search for those seeds occupies most of this bird's time. It is often called the "wild canary" because of the male's distinctive yellow breeding plumage. During the winter, however, that color fades to a dull olive yellow. Goldfinches have a strong attraction to thistledown, which they collect and use to line their nests. Nestlings dine on partially digested seeds regurgitated by the parents. This bird is noted for its constant melodious chatter and is most common in Kansas in the winter when numerous migrants arrive from Canada to join those that permanently reside in our state. *Bob Gress*

There are two kinds of GRAY TREEFROGS in Kansas, both found only in the eastern third of the state. Although they are genetically distinct (one has 12 pairs of chromosomes and the other 24), they look exactly alike, and nobody can tell them apart, except the frogs themselves. Some say their calls are different, and this may be true, but frog calls vary with air temperature. Others say they each live in different habitats, but this is in dispute. All of this is probably a huge joke to the treefrogs, who apparently have no trouble at all sorting each other out. On top of the confusion over their calls and where they live, both species can change color like chameleons, going from gray to bright green and back and displaying varying patterns on their backs and legs. However, if you catch a frog with big pads on the tips of its toes in Kansas, you can be sure you have some kind of Gray Treefrog. Even if it's green, it's a Gray Treefrog. Wait a while, you'll see. Don't bother to check the chromosomes.
Suzanne L. Collins and Joseph T. Collins

The Northern Bobwhite is
found in every county in Kansas
and is a permanent resident of our
state. These birds, also known as
quail, lay clutches of 12 to 14 eggs in
early to mid-April. Studies indicate
that most Bobwhite have a high
mortality rate and rarely live long
enough to breed more than once.
This bird is more plentiful in eastern Kansas because of its preference for forest-meadow habitat
and timber areas that border grain
fields. In western Kansas it is found
mostly in brushy areas along
streams. The Northern Bobwhite is
the prime bird hunted for sport in
our state, with a harvest of up to 2
million birds reported in some
years. *Gerald J. Wiens*

The KILLDEER is well known for its "broken-wing" antics used to lure predators away from its nest site. This cautious bird always nests in open barren areas that provide good vision in all directions. Its spotted eggs, lying in a shallow rocky depression, are difficult to see, and if the nest is approached too closely, the Killdeer goes into its act. This shorebird is at home great distances from water, and its habit of nesting on flat rocky areas has resulted in some nests being built on gravel rooftops. The primary food of this bird is insects, although it also eats many kinds of aquatic invertebrates. *Bob Gress*

The tiny LINCOLN'S SPARROW is a winter visitor to Kansas. It was named by John James Audubon in honor of Thomas Lincoln, a young man who accompanied Audubon on his Labrador trip of 1833. A very secretive bird, this sparrow spends much of its time skulking through underbrush, especially along waterways. When searching for food, it scratches in the leaves on the ground by kicking back with both feet, at the same time maintaining enough balance to avoid falling on its beak. Lincoln's Sparrows spend their summers to the north, staying cool from New England through Canada to Alaska. *Gerald J. Wiens*

The DARK-EYED JUNCO, a member of the sparrow family, is strictly a winter visitor to Kansas, and hence is often called the "snow bird." The name junco is derived from the Latin word *juncus,* meaning seed. In 1973, the American Ornithological Union, a sort of Supreme Court for deciding the common names of birds, lumped 4 species of juncos into a single new species. Two of the former species, now considered subspecies, are commonly found in Kansas. The photograph is of a Slate-colored Junco, the most common variant found in our state. Juncos are found in large flocks during the winter and are primarily ground feeders, spending their time searching for exposed seeds. They usually arrive in Kansas during October and stay until April. *Bob Gress*

The Red Bat is one of the few mammals in which coloration differs between sexes. The male is bright red or orange-red, and the female tends to be a dull red or chestnut. The photograph is of a mother and her two offspring. Although juveniles are not carried in flight, female Red Bats with adolescents attached are sometimes found on the ground, blown there by high winds or frightened from their roost by danger and unable to fly due to the weight of the young. After the young bats leave the roost for good, this creature becomes solitary, hanging 4 to 10 feet above ground by day, hidden in dense foliage that provides shade from above but is open below, allowing a downward fall into flight. Like most Kansas bats, this species migrates south to enjoy warmer weather during the winter months. It feeds primarily on insects such as moths and beetles and often searches for these morsels at night around street lights. *Bob Gress*

Unique among North American mammals in having an armored shell of plates, the Nine-banded Armadillo often leaps straight into the air when startled while crossing a busy highway, thereby making useless its stout covering and placing its entire being in the path of an oncoming vehicle's bumper or undercarriage, with predictable results. The Armadillo has been used in medical research because it is one of the few creatures, besides people, that contracts leprosy. In addition, females always give birth to identical quadruplets, a phenomenon useful in experimental procedures. The Nine-banded Armadillo was virtually unknown north of the Rio Grande valley prior to 1870. Today, this armored immigrant is found throughout much of the southeastern United States and as far north as the Kansas-Nebraska border. *Gerald J. Wiens*

Like its western counterpart, the larger EASTERN HOGNOSE SNAKE also has a turned-up snout as well as a similar tendency to play 'possum when danger looms. The snake in this photograph is newly hatched and more brightly colored than an adult, which is normally patterned with duller shades of brown, black, and yellow. Eastern Hognose Snakes prefer to live along rivers and streams, particularly in sandy areas where toads are abundant. Unlike its western relative, which eats anything it can find, the Eastern feeds exclusively on toads. This reptile has been designated a threatened species in Kansas. It is common in the eastern part of our state but is scarce to the west where water and toads are in limited supply. *Suzanne L. Collins and Joseph T. Collins*

The Blue Jay is a saucy, bold bird that heralds its presence to any within earshot and scares the wits out of smaller birds at the backyard feeder. Its striking pattern and color make this mischievous creature easy to identify. Jays are the self appointed sentinels of the bird world. When a Blue Jay spots an owl, it sets up a raucous chatter, calling in all other Blue Jays from the surrounding neighborhood, whereupon they form a screeching mob and harass the owl into an undignified retreat. The feeding habits of this bird have given it a bad reputation, probably perpetuated by John James Audubon's early-nineteenth-century drawing of a Blue Jay robbing eggs from a nest. In reality, eggs and nestlings are a minor portion of this bird's diet, whereas nuts, berries, and seeds make up the majority of its food. *Gerald J. Wiens*

A permanent resident of eastern Kansas, the Tufted Titmouse is a bird that seldom sits still. Bigger than its close relatives, the Chickadees, this crested bird raids a backyard feeder with the same impunity as its smaller cousins, and displays their same acrobatic maneuvers, sometimes hanging upside down to dine on suet. Although it feeds on suet and sunflower seeds during the winter months, it eats insects all year long, catching them in the summer and digging them out of the holes and crevices of bark in winter. The Tufted Titmouse nests in the cavities of tree trunks and large limbs, where it lays its eggs and raises its young. *Bob Gress*

The Carolina Chickadee, a quick, alert little bird, is found mostly in southeastern Kansas and closely resembles its more wide-ranging Black-capped cousin found throughout the rest of our state. Chickadees are a delight to watch, racing about in small groups in search of seeds and other good things to eat. Their arrival at a bird-feeder is like an invasion of madcap marauders—the other birds at the feeder seem momentarily taken aback by these audacious creatures as they dart in, snatch food, and race off to a nearby perch. Chickadees are permanent residents of Kansas, with both of the species being more abundant in the eastern part. *Bob Gress*

Early settlers to our country called the EASTERN BLUEBIRD the "blue robin" because of its red breast. Like the true Robin, it is a member of the thrush family. When cavity-nesting House Sparrows and European Starlings were imported from the Old World to our nation, both entered into competition with the Eastern Bluebird for places in which to lay their eggs and raise young. Such pressure from alien species, coupled with insecticide use, destruction of nesting habitat, and severe weather, has caused this beneficial, insect-eating bird to experience a population decline. In hopes of speeding its recovery, Bluebird nest boxes like the one in this photograph are being built and installed throughout the range of this creature. *Bob Gress*

CEDAR WAXWINGS arrive in Kansas each winter from the north and are often first noticed because of their high-pitched calls which are used to coordinate flock movements. Cedar Waxwings spend a lot of time in cedar trees, greedily gulping down the sky-blue berries until they are so gorged and heavy that takeoff is a moment of uncertainty. The bright red tips on this bird's secondary wing feathers exude waxlike droplets, whose function is unknown. With its silky feathers, head crest, and black mask, the sleek Cedar Waxwing is a true nomad, visiting our state unexpectedly and leaving just as abruptly. Its unpredictable travels are probably based on availability of food. A male Waxwing sometimes presents a flower petal to his mate, an offering or gift that probably goes a long way in settling any avian disputes of the couple.
Bob Gress

In this RACCOON family, the mother (the big one on the right in the photograph) was raised in captivity and then released. The litter shown in the photograph consisted of 7 babies, but one was too shy to descend from the safety of its tree. Raccoons are expert climbers and prefer to take refuge in trees or in water. They have sensitive front feet that they use to feel around underneath logs and rocks for food. Contrary to popular belief, they don't wash their food before eating. However, much of what they like to eat lives in water, so naturally they spend a lot of time with their front paws feeling around beneath the water's surface. Furthermore, they sometimes like to dunk a hard bit of food into water in order to soften it. *Gerald J. Wiens*

Unlike most other birds, the MISSISSIPPI KITE will vigorously defend its nest and young from people, swooping from the sky to slap and swipe at the person unlucky enough to get too near. Rarely, however, do the birds actually make contact. They are called Kites because of their ability to soar and glide with little apparent wing movement, as though suspended at the end of a string. Their favorite food appears to be cicadas, and they are quite agile in their pursuit of these insects, plucking the unlucky morsels from the trunks of trees while in full flight. *Bob Gress*

Formerly called the "marsh hawk," the NORTHERN HARRIER lives in open prairies and prairie marshes. Its flight is a characteristic swift glide as it flies close to the ground, displaying its distinctive white rump. The Northern Harrier is primarily a winter visitor to Kansas, but some of these birds also nest here. This is a slender bird when compared to other raptors. It feeds on rats, mice, frogs, and snakes and has been observed following prairie fires looking for rodents as they scurry to escape the flames. *Gerald J. Wiens*

SNOW GEESE come in two variants, one dark with a white head and the other all white, both of which are shown in the photograph. The dark version is sometimes called the "Blue Goose," but both are the same species, and the more widely accepted name is Snow Goose. The two color variations were once considered to be two distinct species; but both are now known to freely interbreed, and the young of both kinds can emerge from the same clutch. Snow Geese migrate in a narrow corridor along the eastern edge of our state, headed for the Gulf of Mexico. An excellent place to observe these big birds in Kansas is at Brown County State Lake in mid-October. *Bob Gress*

In summer, the WOOD DUCK is a common resident in eastern Kansas but is not so abundant in the west. That it exists at all is one of the major success stories of this century. Hunting and habitat destruction had devastated this bird by the turn of the century, but due to federal protection enacted in 1918 it has recovered throughout much of its range. The male in this photograph is strikingly colored and patterned, unlike the two females who are drab by comparison. Wood Ducks nest in tree cavities in forests along our rivers and streams. As many as a dozen ducklings eventually jump from the nest, which may be 40 feet up, and bounce harmlessly on the ground. Once on the water, young Wood Ducks are frequently consumed by turtles or large fishes. Luckily, enough of them survive to grace our waterways, nest in our woods, and remind us that with education and concern we can save our wildlife and sustain the diversity of our environment. *Gerald J. Wiens*

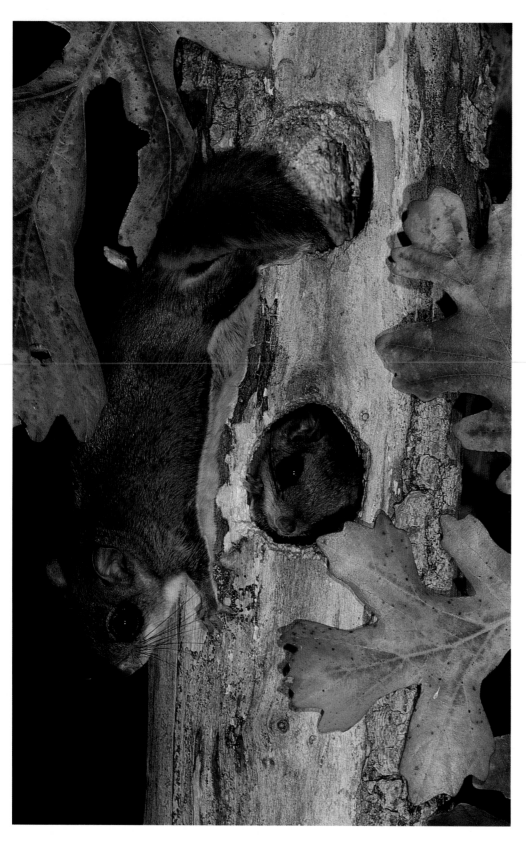

The Southern Flying Squirrel (left) is the smallest of Kansas squirrels (its body is only 4 to 5 inches long) and is restricted to the heavily wooded areas of the eastern part of our state. Even when present in an area, these big-eyed, delicate squirrels are seldom seen because they are nocturnal and very secretive. Their flattened tails are used as rudders as they jump and glide from tree to tree. They often nest in woodpecker holes, emerging at night to search for nuts, seeds, berries, buds from trees, and sometimes insects. *Bob Gress*

The Fox Squirrel (right) gets its name from the color of its reddish hair, which gives it a foxlike color. This largest of North American tree squirrels is familiar to most Kansans who have bird feeders because these mammals are extremely persistent and quite acrobatic in raiding a feeder. They sometimes make large leafnests in trees, which may seem a cold way to live, but when all curled up inside with their bushy tails, they manage to stay warm. Most Fox Squirrels, however, prefer tree cavities instead of exposed leafnests. They feed on all kinds of nuts, including acorns, pecans, walnuts, and hickory nuts, and also consume fruits, berries, and tender buds. *Bob Gress*

The male NORTHERN CARDINAL can be quite solicitous of his mate, bringing her food during courtship and incubation and caring for their first brood while she is incubating the second. The word "cardinal," which originally meant "important," evolved to refer to high-ranking members of the Roman Catholic Church and more recently has become associated with the bright red robes worn by those church officials. This brightly vested bird was named for the latter. Cardinals are grosbeaks and relish sunflower seeds but also like to eat insects and fruit. The resounding, cheerful call of this bird is easily recognized, even though it is known to consist of at least 28 different songs. Both males and females sing, sometimes together, which is unusual in the avian world. *Gerald J. Wiens*

The EVENING GROSBEAK was once mistakenly thought to sing mainly in the evening. We now know it sings all day long, but nonetheless the name stuck. Evening Grosbeaks are members of the finch family and get the name "grosbeak" because of their large, or "gross," conical-shaped bills. The brightly colored male in this photograph is quite different from the female, who wears a coat of silver-gray with just enough yellow, black, and white to tell she's a member of the same species. This is a bird of the north; most members of the species spend the entire year in New England, southern Canada, and the mountains of the western United States. A few drop down to Kansas during the winter months and are easily attracted to backyard feeders with sunflower seeds. *Gerald J. Wiens*

One of the most variably patterned snakes in our state, the GROUND SNAKE may be completely banded like the one in this photograph, partly banded, or not banded at all. This is a reptile of the southern part of our state, although an isolated population exists near Hays. It is most abundant in the Red Hills of south-central Kansas along the Oklahoma border where it lives beneath rocks and in the cracks and crevices of the dry, red soil that covers the ravines and gullies of that region. This is a small snake, reaching a maximum length of about 15 inches. It feeds on spiders, scorpions, centipedes, and other tasty items. *Suzanne L. Collins and Joseph T. Collins*

The BULLSNAKE is the biggest snake in Kansas, with the record 88¾-inch individual tipping the scales at nearly 8 pounds. This is also the most economically beneficial snake in our state because it consumes huge quantities of rats and mice, thereby preventing the loss of stored grain. Bullsnakes are abundant statewide but prefer the open prairies and plains to the forests of our eastern border. Females lay their clutches in soft earth beneath logs or large rocks, and the 5 to 19 eggs hatch in about 2 months. Like the Eastern and Western Hognose Snakes, this reptile can emit a hiss loud enough to scare the tar out of you if you are unaware of the snake's presence. *Suzanne L. Collins and Joseph T. Collins*

Probably the most brightly colored and attractive snake in Kansas, the RED MILK SNAKE is much feared by many Kansans because they mistake it for the venomous Coral Snake. This could not be true, of course, because Coral Snakes are not found in our state. But this reptile has even more going against it. Besides its close resemblance to a dangerous reptile, it has the word "milk" in its name, the implication being that it likes milk, and this probably doesn't endear it to dairy farmers. However, just as the Red Milk Snake is not venomous, this imitation candycane also does not drink milk. Its common name probably was earned because it sometimes hangs out around cow-barns in search of mice and lizards, which it does like to eat. Milk Snakes are found throughout Kansas. *Suzanne L. Collins and Joseph T. Collins*

The KANSAS GLOSSY SNAKE lives in the sand dunes and loose soils of southwestern Kansas, ranging as far north along the Colorado border as Cheyenne County and east along the Oklahoma border to Sumner County. It is completely nocturnal, moving about in search of its favorite food, lizards. Apparently its name comes from its glossy appearance in the beam of a flashlight or the headlights of a car. Kansas Glossy Snakes are egg-layers and annually produce clutches of up to 23 eggs. The eggs require 2 or 3 months to hatch, depending on daily temperature during incubation. This docile reptile has been designated a threatened species in Kansas. *Suzanne L. Collins and Joseph T. Collins*

The BOBCAT is very shy and difficult to approach and observe because of its keen sense of smell and excellent hearing. It eats mostly rabbits, rodents, and birds. This diet of readily available and abundant prey plus the Bobcat's small size and reclusive habits probably explains why this feline still thrives throughout Kansas today. Besides people, it has few enemies; histori-cally its only major predator was the Mountain Lion, an extirpated species whose recent return to Kansas has been rumored, but not yet confirmed. Bobcats den in rock crevices, hollow logs, and windfalls, where the female raises an average litter of 3 kittens. This mammal gets its name from its short, bobbed tail. *Bob Gress*

Although sometimes called the American antelope, the graceful PRONGHORN is not a close relative of the true antelopes of Africa. It is the fastest mammal in the Western Hemisphere and one of the fastest in the world. A Pronghorn can bound 20 feet and can reach and sustain 70 mph for 3 to 4 minutes. Bursts of 45 mph are not unusual, and it can easily cruise at 30 mph for up to 15 miles. It runs with its mouth open, not from exhaustion, but to gasp extra oxygen. Pronghorns inhabit open terrain and rely on their eyesight and speed to avoid danger. Their large eyes have a wide arc of vision and can detect movement 4 miles away. If alarmed, the Pronghorn raises the white hairs on its rump, creating a flash of white visible at great distances. If forced to fight, this animal will use its sharp hooves, which are effective enough to drive off a Coyote. The short prongs on its horns give this animal its common name. Unlike deer antlers, which are shed each year, only a thin horny sheath is shed annually by the Pronghorn. *Gerald J. Wiens*

The official state mammal of Kansas, historically, the Bison was the most important animal on the Great Plains, providing food and clothing for American Indians. It once roamed the prairies in massive herds, migrating to and from feeding areas over extended periods of time. In the 1800s it was hunted mercilessly and driven to the edge of extinction. Today, Bison exist in Kansas only as small, managed herds held in enclosures for public viewing, a sad testament to our nation's abuse of wildlife.
Gerald J. Wiens

SUGGESTED REFERENCES

Bee, James W., Gregory E. Glass, Robert S. Hoffmann, and Robert R. Patterson. 1981. *Mammals in Kansas.* University of Kansas Museum of Natural History Public Education Series No. 7. 300 pp.

Burt, William Henry. 1976. *A Field Guide to the Mammals of North America North of Mexico.* Third Edition. Peterson Field Guide No. 5. Houghton Mifflin Company (Boston). 289 pp.

Caldwell, Janalee P., and Joseph T. Collins. 1981. *Turtles in Kansas.* AMS Publishing (Lawrence, Kansas). 67 pp.

Collins, Joseph T. 1982. *Amphibians and Reptiles in Kansas.* Second Edition. University of Kansas Museum of Natural History Public Education Series No. 8. 356 pp.

Collins, Joseph T. (ed.). 1985. *Natural Kansas.* University Press of Kansas (Lawrence). 226 pp.

Collins, Joseph T., and Suzanne L. Collins. 1991. *Reptiles and Amphibians of the Cimarron National Grasslands, Morton County, Kansas.* U. S. Forest Service (Elkhart, Kansas). 60 pp.

Conant, Roger, and Joseph T. Collins. 1991. *A Field Guide to the Reptiles and Amphibians of Eastern and Central North America.* Third Edition. Peterson Field Guide No. 12. Houghton Mifflin Company (Boston). 450 pp.

Cross, Frank B., and Joseph T. Collins. 1975. *Fishes in Kansas.* University of Kansas Museum of Natural History Public Education Series No. 5. 189 pp.

Johnston, Richard F. 1965. *A Directory to the Birds of Kansas.* University of Kansas Museum of Natural History Miscellaneous Publication No. 41. 67 pp.

Page, Lawrence M., and Brooks M. Burr. 1991. *A Field Guide to the Freshwater Fishes of North America North of Mexico.* Peterson Field Guide No. 42. Houghton Mifflin Company (Boston). 432 pp.

Peterson, Roger Tory. 1980. *A Field Guide to the Birds of Eastern and Central North America.* Fourth Edition. Peterson Field Guide No. 1. Houghton Mifflin Company (Boston). 384 pp.

Thompson, Max C., and Charles Ely. 1989. *Birds in Kansas.* Volume One. University of Kansas Museum of Natural History Public Education Series No. 11. 404 pp.

Tomelleri, Joseph R., and Mark E. Eberle. 1990. *Fishes of the Central United States.* University Press of Kansas (Lawrence). 226 pp.

Zimmerman, John L., and Sebastian Patti. 1988. *A Guide to Bird Finding in Kansas and Western Missouri.* University Press of Kansas (Lawrence). 244 pp.

ABOUT THE PHOTOGRAPHERS

BOB GRESS is the naturalist/director of a Wichita Park Department nature education program, Wichita Wild. His photographs have appeared in calendars, postcards, books, and magazines, including *National Wildlife, Birder's World, American Birds, Birding, Kansas! Kansas Wildlife and Parks, Birds in Kansas,* and *Natural Kansas.* His photographs have received several awards, including first-place finishes in the 1990 *National Wildlife* magazine photography contest and the 1990 *Wildbird* magazine photography contest, and twice have won the Kansas Nongame Wildlife photography contest sponsored by the Department of Wildlife and Parks. His photographs of live animals have illustrated over 2,000 wildlife programs. He was honored by the Kansas Wildlife Federation as the 1989 Conservation Educator of the Year and was the recipient of the City of Wichita's 1990 Excellence in Public Service Award.

GERALD J. WIENS is a naturalist who has integrated his talents as a wildlife photographer into the production of nature education programs. His photographs have been winners in the National Wildlife Federation wildlife photography contest, Kansas Nongame Wildlife photography contest, and the *Sports Afield* photography contest. He has been a regular contributor to *Kansas!* magazine, and has had photographs in *Weekly Reader, World of Wildlife Calendar, In Your Big Backyard, Kansas Wildlife,* and *Camera Press, Ltd.* of London. Books that contain his photographs include *Natural Kansas, Birds in Kansas,* and the National Wildlife Federation's *December Treasury.*

SUZANNE L. COLLINS is a wildlife photographer with credits in numerous magazines and journals, and in such books as *The American Milk Snake, Amphibians of Oklahoma,* and *Natural Kansas.* A portfolio of her work entitled "Amphibians in Kansas" is on display at the Museum of Natural History, the University of Kansas. She and Joseph T. Collins were the 1990 winners of the Kansas Nongame Wildlife photography contest. She coauthored (with Joseph T. Collins) *Reptiles and Amphibians of the Cimarron National Grasslands, Morton County, Kansas,* a book that includes forty of her color photographs. She is an assistant to the dean at the University of Kansas School of Education.

JOSEPH T. COLLINS is the state's most prolific book author on the subject of Kansas wildlife. His titles include *Amphibians and Reptiles in Kansas* (two editions), *Fishes in Kansas* (with Frank B. Cross), *Turtles in Kansas* (with Janalee Caldwell), and *Natural Kansas.* He also edited *Mammals in Kansas* and *Birds in Kansas, Volume One,* and is coauthor (with Roger Conant) of the third edition of the *Peterson Field Guide to Reptiles and Amphibians of Eastern and Central North America.* He is a zoologist and editor at the Museum of Natural History, the University of Kansas.

INDEX TO PORTRAITS

Library of Congress Cataloging-in-Publication Data

Collins, Joseph T.
    Kansas wildlife / text by Joseph T. Collins ; with photographs by Bob Gress ... [et al.] ;
foreword by John E. Hayes, Jr.
       p. cm.
    Includes bibliographical references and index.
    ISBN 0-7006-0503-7 (alk. paper)
    1. Vertebrates—Kansas—Pictorial works. I. Gress, Bob.
II. Title.
QL606.52.U6C65 1991
596.09781—dc20   91-18859

Printed in Singapore

10 9 8 7 6 5 4 3 2

This book is printed on acid-free paper.

Design by John Baxter